Earth's
LAYERS

Rebecca Woodbury, Ph.D., M.Ed.

Gravitas Publications Inc.

Earth's Layers

Illustrations: Janet Moneymaker

Earth's Layers
ISBN 978-1-950415-33-5

Published by Gravitas Publications Inc.
Imprint: Real Science-4-Kids
www.gravitaspublications.com
www.realscience4kids.com

RS4K

Photo credits: Cover & Title Pg: Giuseppe Dio from Pixabay; P 3. Image by Pexels from Pixabay;
P.5. Lukas, AdobeStock; P.17. Jacob Lawler on Unsplash;

Earth is called a
terrestrial planet.
Terrestrial planets are
made mostly of rock.

Rocks are
everywhere!

We can see that Earth is made of rocks, dirt, oceans, and large pieces of land.

But what is beneath the rocks, dirt, oceans, and land that we see on the outside of Earth?

Scientists think that if Earth could be cut in half, we would see at least three **layers**.

Cakes can have layers too!

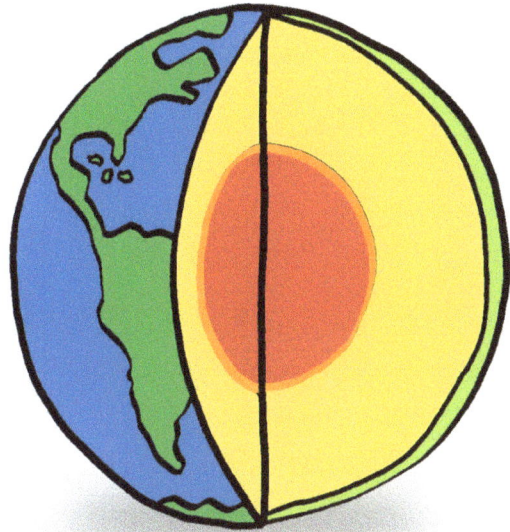

The outer layer of Earth is called the **crust.** The crust is made of rocks, minerals, and soil.

Is it like pie crust?

No. It's not that kind of crust.

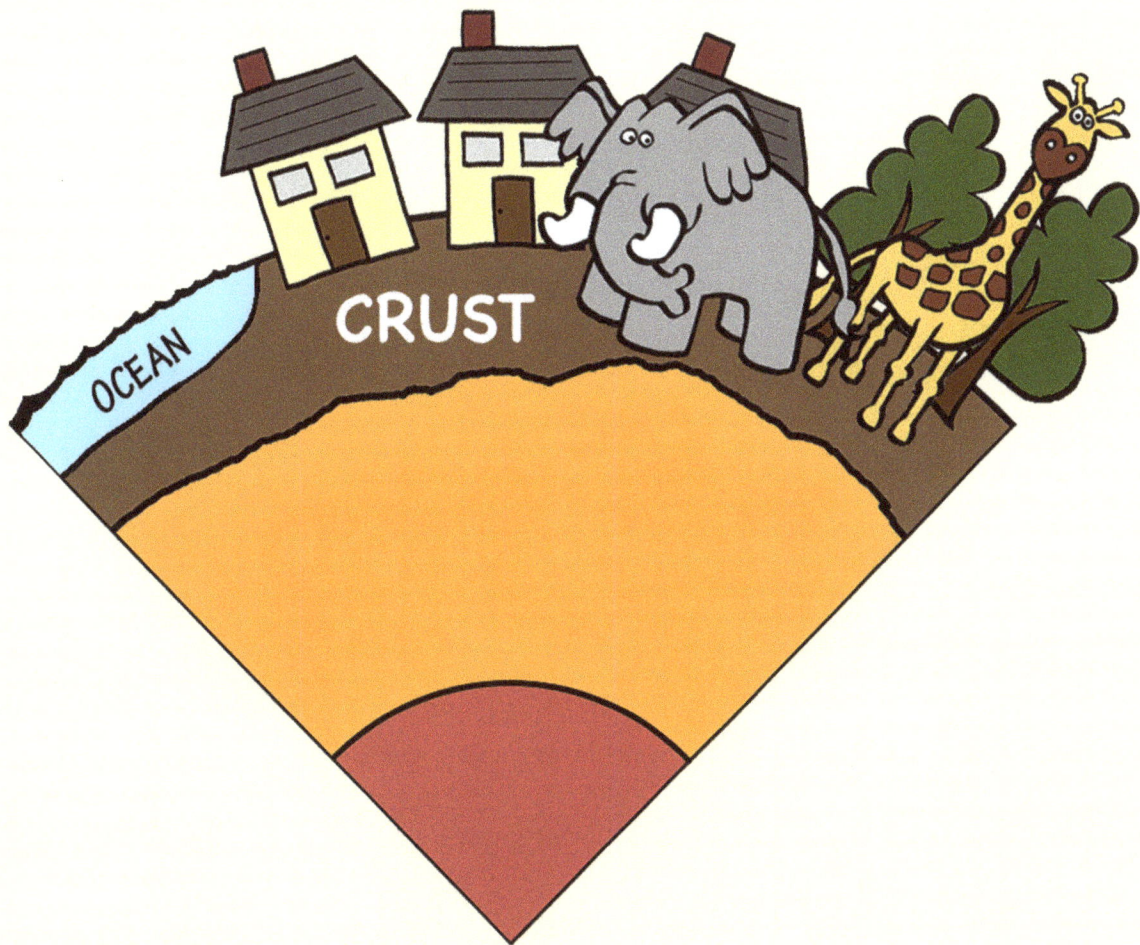

OCEAN

CRUST

Below the crust is the **mantle.**
Scientists think the top part of the
mantle is hard and rocky and
the lower part is soft like peanut
butter. The mantle is thought to
be thicker than the crust.

I LOVE peanut butter!

The mantle is not **really** peanut butter!

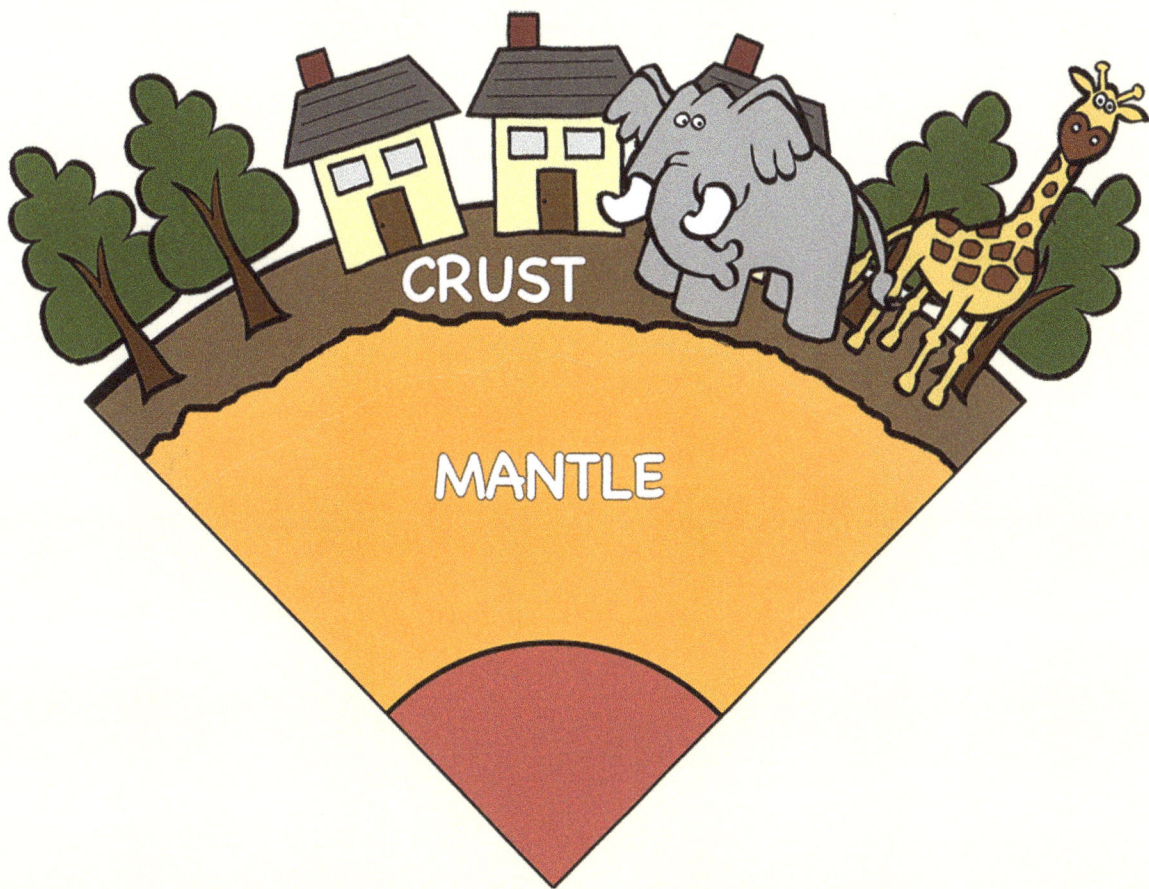

CRUST

MANTLE

In the center of Earth is the **core**. Scientists think that the core is a ball made of **iron** and **nickel.** The core is most likely soft on the outside and solid on the inside.

Is it like an apple core?

Silly mouse!

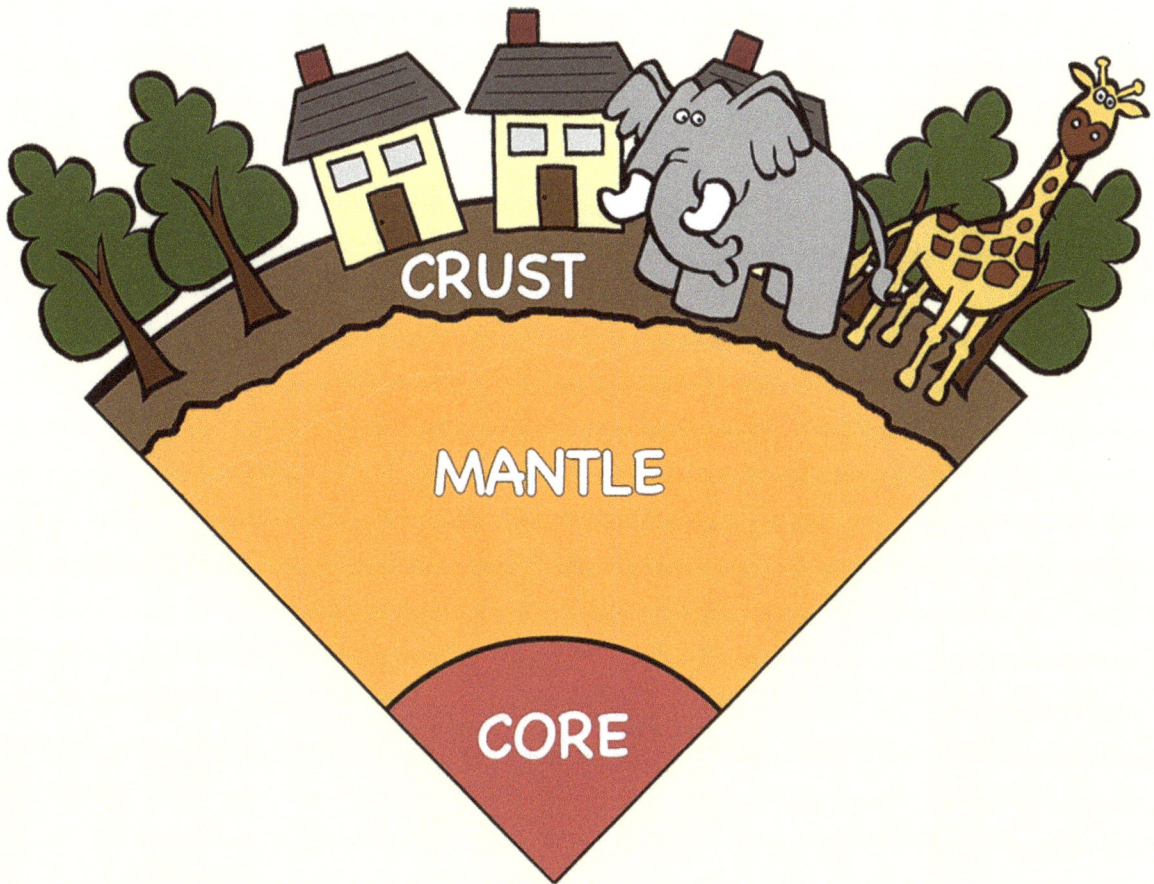

CRUST

MANTLE

CORE

Scientists cannot actually see what is deep below the Earth's surface. But they can test ideas about what might be there by studying things like **volcanoes** and **earthquakes**.

A volcano!

During a volcano hot, melted material called **magma** pushes through the crust. Seeing and studying volcanoes and magma gives scientists an idea of what is below the crust.

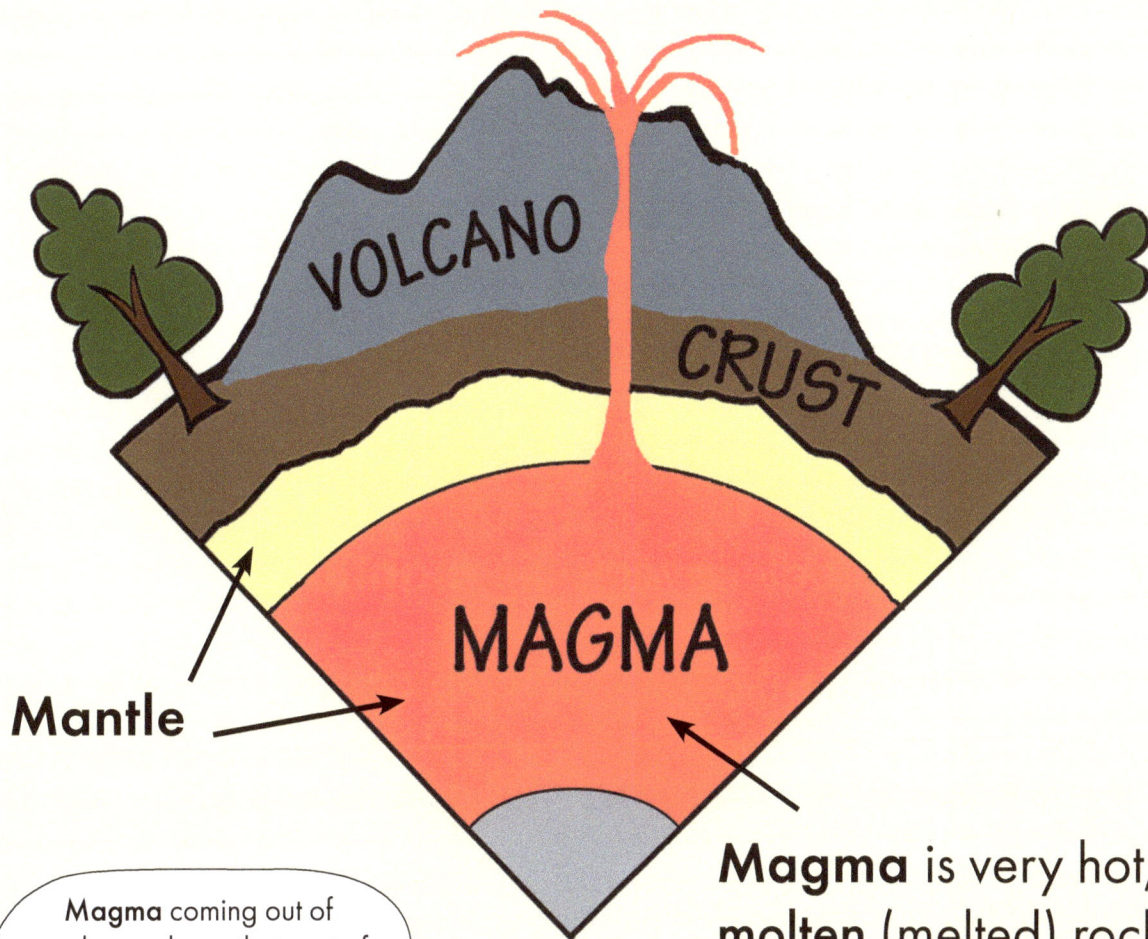

VOLCANO

CRUST

MAGMA

Mantle

Magma is very hot, **molten** (melted) rocks.

Magma coming out of a volcano shows that part of the mantle is most likely soft!

SUMMARY

- Earth has **layers.**

- The outer layer is the **crust.**

- The middle layer is the **mantle.**

- The innermost layer is the **core.**

- Studying **magma** that comes out of **volcanoes** can help us learn what Earth is like below the crust.

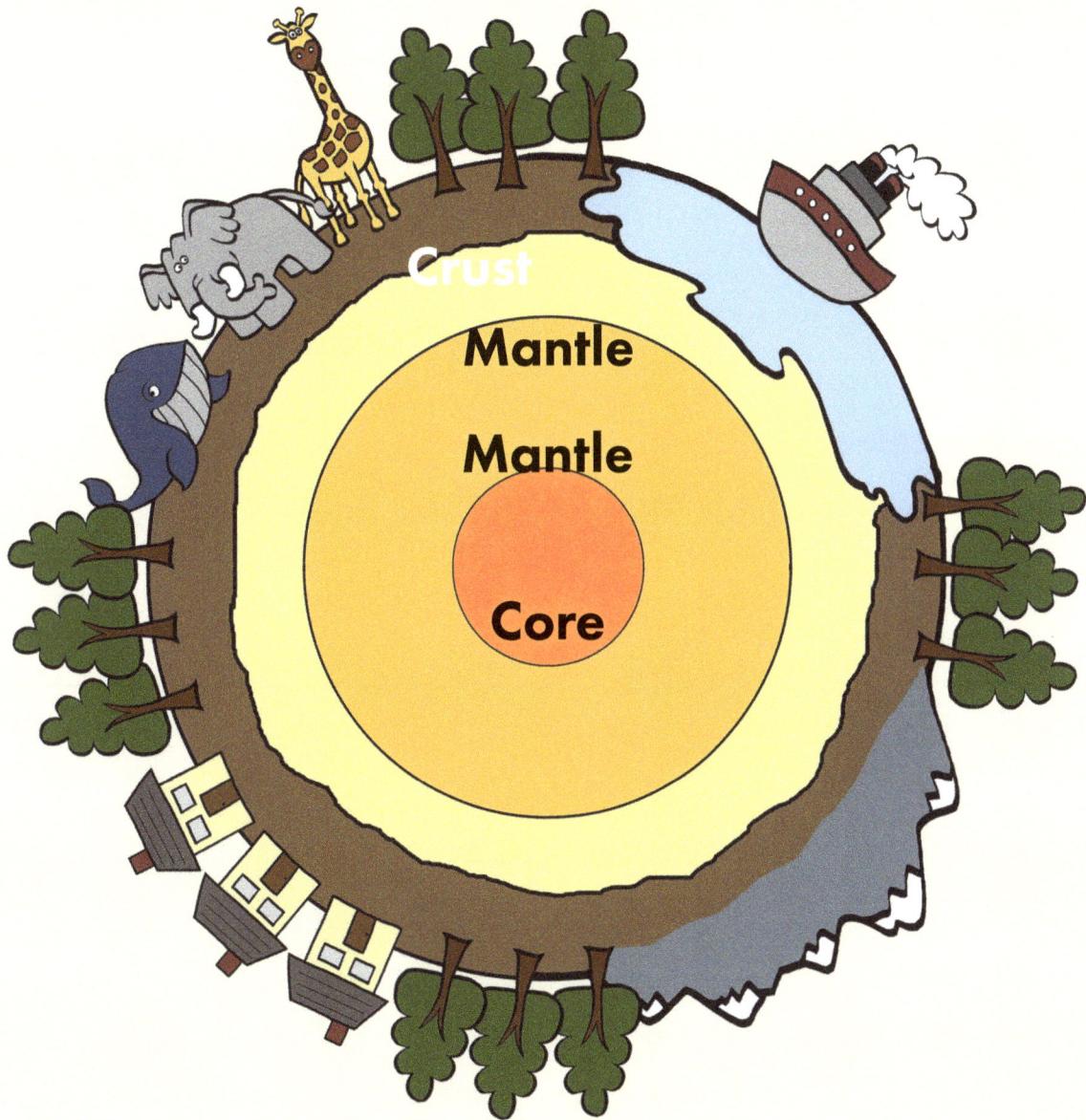

Crust

Mantle

Mantle

Core

How to say science words

core   (KAWR)

crust   (KRUHST)

Earth   (ERTH)

earthquake   (ERTH-kwayk)

geologist   (jee-AH-luh-jist)

iron   (IY-uhrn)

layer   (LAY-uhr)

magma   (MAAG-muh)

mantle   (MAAN-tuhl)

mineral   (MIHwN-ruhl)

nickel   (NIH-kuhl)

planet   (PLAA-nuht)

scientist   (SIY-uhn-tist)

terrestrial   (tuh-RESS-tree-uhl)

volcano   (vahl-KAY-noh)

www.ingramcontent.com/pod-product-compliance
Lightning Source LLC
Chambersburg PA
CBHW040149200326
41520CB00028B/7544